MONSTERS DO MATH!

BIGFOOT FINDS FRACTIONS!

BY THERESE M. SHEA

Gareth Stevens

Please visit our website, www.garethstevens.com. For a free color catalog of all our high-quality books, call toll free 1-800-542-2595 or fax 1-877-542-2596.

Cataloging-in-Publication Data

Names: Shea, Therese M.
Title: Bigfoot finds fractions! / Therese M. Shea.
Description: New York : Gareth Stevens Publishing, 2021. | Series: Monsters do math! | Includes glossary and index.
Identifiers: ISBN 9781538257197 (pbk.) | ISBN 9781538257210 (library bound) | ISBN 9781538257203 (6 pack)
Subjects: LCSH: Fractions--Juvenile literature. | Mathematics--Juvenile literature. | Sasquatch--Juvenile literature.
Classification: LCC QA117.S54 2021 | DDC 513.2'6--dc23

First Edition

Published in 2021 by
Gareth Stevens Publishing
111 East 14th Street, Suite 349
New York, NY 10003

Designer: Sarah Liddell
Editor: Char Light
Illustrator: Bobby Griffiths

Photo credits: p. 4 Aleksandra H. Kossowska/Shutterstock.com; p. 7 andreiuc88/Shutterstock.com; p. 9 (background) Evgeny Karandaev/Shutterstock.com; p. 9 (granola bar) baibaz/Shutterstock.com; p. 10, 12 Bettmann/Contributor/Bettmann/Getty Images; p. 16 Sharon Keating/Shutterstock.com; p. 18 Clayton D. Will/Shutterstock.com; p. 20 Travis Wolfe/Shutterstock.com.

Printed in the United States of America

CPSIA compliance information: Batch #CS20GS: For further information contact Gareth Stevens, New York, New York at 1-800-542-2595.

CONTENTS

Words in the glossary appear in **bold** type the first time they are used in the text.

BIGFOOT SPOTTED!

Do you hear that? Something is walking through the woods in the park. Maybe it's a bear! Do you smell that? It smells bad. Maybe it's a skunk! Do you see that? It's not a skunk or a bear. It's like a man, but much larger and much hairier. It's Bigfoot!

People likely won't believe you saw Bigfoot. He's known for being shy. He's come out of the woods to teach you about fractions. Get ready to learn from the wildest teacher you'll ever have! Check your answers to the questions in this book on page 22.

NO ONE HAS PROVEN THAT BIGFOOT EXISTS—OR DOESN'T EXIST—YET!

EQUAL PARTS OF A WHOLE

Bigfoot sees fractions everywhere. He uses a stick to draw circles on the ground. He **divides** them into equal parts. "Fractions!" Bigfoot **growls.** A fraction is a part of a whole. Bigfoot is showing you a whole can be broken up equally in different ways.

SOME PEOPLE THINK CREATURES LIKE BIGFOOT COULD BE AN EARLY FORM OF HUMANS.

A fraction's bottom number is the number of equal parts that make up the whole. The top number **represents** some of those parts. In this case, it represents the number of colored parts. In the fraction $\frac{3}{4}$, how many equal parts make up the whole?

MONSTER FACTS!

BIGFOOT'S OTHER NAME, SASQUATCH, COMES FROM A WORD IN THE LANGUAGE OF THE SALISH NATIVE AMERICANS. IT MEANS "WILD MEN."

FINDING A FRIEND

Bigfoot **gestures** to you. He's telling you to follow him. You go deeper into the woods. He points to a cave and growls, "Home!" Another Bigfoot comes out. You've discovered not only that Bigfoot exists, but that there's more than one!

THREE-THIRDS, OR 3/3, IS THE SAME AS THE WHOLE, OR 1.

By now, you're hungry. You have a **granola** bar. You give it to Bigfoot. He takes a sharp rock and cuts the bar into 3 equal pieces. He takes 2 pieces for himself and his friend. What fraction of the bar did they leave for you, $\frac{1}{3}$ or $\frac{2}{3}$?

MONSTER FACTS!

BRITISH **EXPLORER** DAVID THOMPSON IS SAID TO HAVE FOUND THE FIRST SET OF BIGFOOT FOOTPRINTS IN 1811.

A YOUNG BIGFOOT

Behind the two Bigfoots, a small Bigfoot comes out of the cave rubbing his eyes. A Bigfoot kid! He must have been sleeping. He smells you—and licks your hand!

MONSTER FACTS!

IN 1967, A MAN NAMED ROGER PATTERSON FILMED, OR RECORDED, WHAT HE SAID WAS A BIGFOOT IN BLUFF CREEK, CALIFORNIA.

ROGER PATTERSON'S FILM MADE MANY PEOPLE BIGFOOT BELIEVERS. OTHERS THINK IT'S A FAKE. A PHOTO FROM THE FILM IS SHOWN HERE.

You look in your bag for something to share with the little Bigfoot. You find a sheet of stickers. You want to divide the stickers equally between you and the little Bigfoot. You share $\frac{1}{2}$, or half, the stickers shown above. How many stickers do you have left?

BIGFOOT PIZZA PARTY!

Maybe it was the noise. Maybe it was the smell. But more Bigfoots come out of the woods to meet you. They bring pizza! You wonder how they got it. Maybe Bigfoots can order pizza too?

MONSTER FACTS!

DR. GROVER KRANTZ WAS ONE OF THE FEW SCIENTISTS WHO BELIEVED IN BIGFOOT. HE SEARCHED FOR **PROOF** FOR MUCH OF HIS LIFE.

DR. GROVER KRANTZ IS SHOWN HERE WITH **CASTS** HE MADE OF HUGE FOOTPRINTS THAT HAD BEEN FOUND.

One pizza is divided into 8 equal parts. Seven pieces are eaten. What fraction of the pizza was eaten, $\frac{7}{8}$ or $\frac{1}{8}$? What fraction of the pizza is left, $\frac{7}{8}$ or $\frac{1}{8}$? Use the picture to help you.

SASQUATCH SOCCER

You can't have a party without games! The Bigfoots play a game that's a bit like soccer. They made a net of leaves and kick balls made out of **twigs** into it. The Bigfoots have 10 balls. They kick 5 balls into the leaf net. What fraction of the balls do they kick in the net?

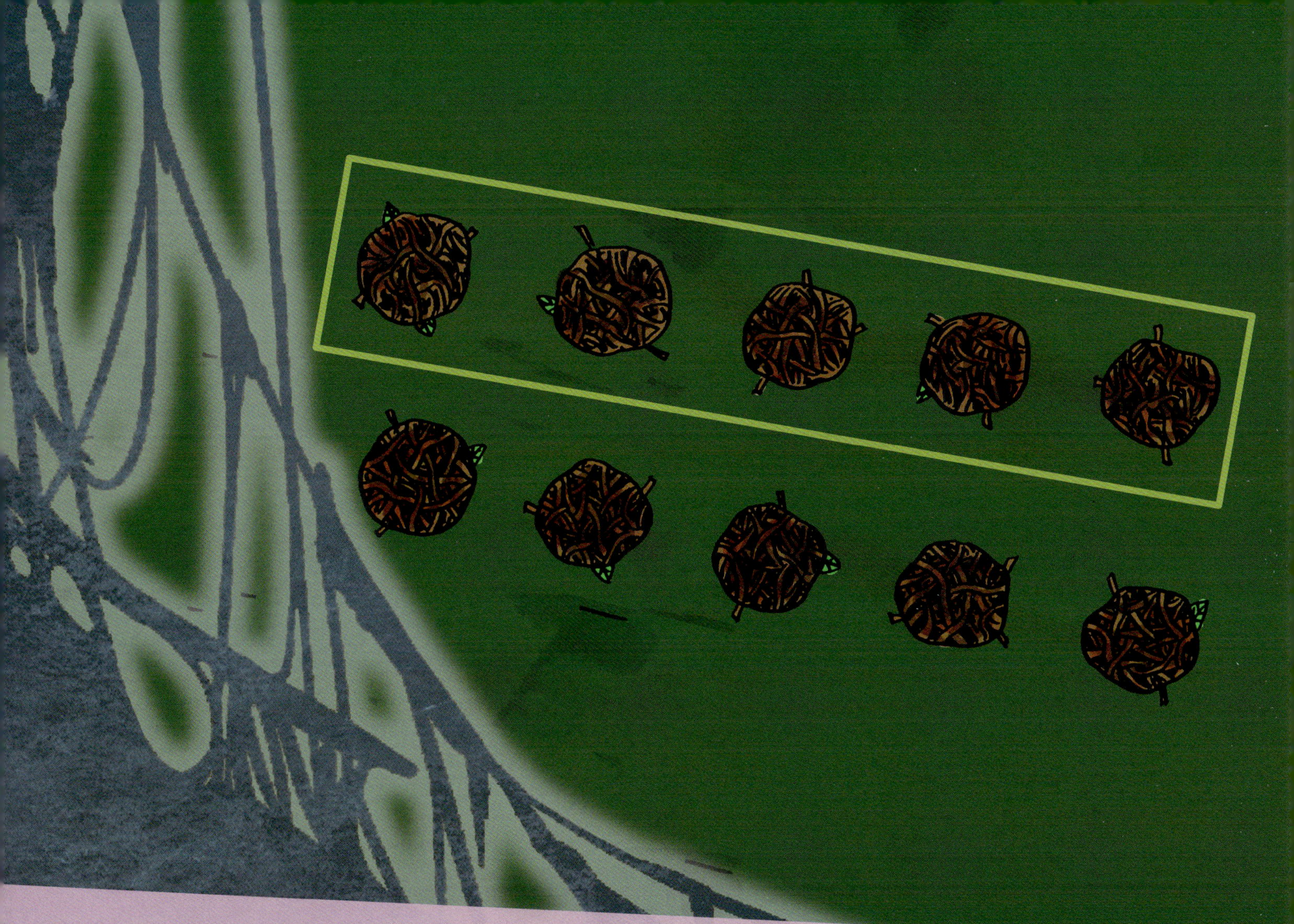

Is the fraction of balls kicked into the net equivalent, or equal, to the fraction $\frac{1}{3}$ or $\frac{1}{2}$?

MONSTER FACTS!

THERE'S A BIGFOOT **FESTIVAL** IN MARION, NORTH CAROLINA. THOUSANDS OF BELIEVERS AND NONBELIEVERS ATTEND.

KNOW ABOUT NUMBER LINES

Bigfoot knows that fractions can seem tricky. He has one more hint about how to understand them better. It's called a number line. He made one and divided it into 6 equal parts. Each part represents $\frac{1}{6}$.

WASHINGTON HAS THE MOST REPORTED BIGFOOT SIGHTINGS OF ANY U.S. STATE.

MONSTER FACTS!

A CREATURE LIKE BIGFOOT HAS BEEN SPOTTED IN THE HIMALAYAS, WHICH ARE MOUNTAINS IN ASIA. THERE, IT'S CALLED THE **ABOMINABLE** SNOWMAN, OR THE YETI.

A number line like this can help you see how two fractions can be equivalent. Find the mark $\frac{2}{6}$ on his number line. Do you think $\frac{2}{6}$ on the number line is equal to $\frac{1}{2}$ or $\frac{1}{3}$? Use the picture to help you answer.

READ THE MAP

It's time to go home. You ask Bigfoot where you are. He shows you a square-shaped map that he made. It's a map of the park. It's divided into equal parts. Bigfoot points out Bigfoot territory and then shows you how to get back to the rest of the park—and home.

MONSTER FACTS!

THERE'S A PATH THROUGH MOUNTAINS IN CALIFORNIA AND OREGON CALLED THE BIGFOOT TRAIL!

How much of the map is Bigfoot territory, $\frac{3}{4}$ or $\frac{1}{4}$? How much of the map is the rest of the park, $\frac{3}{4}$ or $\frac{1}{4}$?

THANKS, BIGFOOT!

Bigfoot brings you out of the woods so you can get home safely. You just need a photo of him so everyone believes you saw him. You ask him to stay put while you run back to the car. But when you return, he's gone! There are some huge footprints left in the dirt though.

The only proof of Bigfoot that you have is your new knowledge of fractions. Keep looking for fractions in the world around you. They're easier to spot than Bigfoot!

NEVER GO LOOKING FOR BIGFOOT BY YOURSELF. ASK AN ADULT TO GO WITH YOU. EVEN IF YOU DON'T FIND HIM, YOU MIGHT FIND FRACTIONS!

GLOSSARY

abominable: describing something unpleasant or disgusting
cast: an object made by pouring melted matter into a mold and letting it harden
divide: to break up
explorer: a person who travels to find scientific information
festival: a special event where people gather to celebrate something
gesture: a movement of the body that shows an idea
granola: a mixture of oats and other foods such as raisins and nuts
growl: to make a deep, low sound
proof: something that shows something else is real or true
represent: to stand for
twig: a small tree branch

ANSWER KEY

p. 7: 4 equal parts
p. 9: $\frac{1}{3}$
p. 11: 6 stickers
p. 13: $\frac{7}{8}$ was eaten; $\frac{1}{8}$ is left
p. 14: $\frac{5}{10}$
p. 15: $\frac{1}{2}$
p. 17: $\frac{1}{3}$
p. 19: $\frac{1}{4}$ is Bigfoot territory; $\frac{3}{4}$ is the rest of the park

FOR MORE INFORMATION

BOOKS

Cole, Bradley. *Bigfoot*. North Mankato, MN: Capstone Press, 2020.

Pistoia, Sara, and Piper Whelan. *Fractions*. New York, NY: Smartbook Media Inc., 2019.

Williams, Zella, and Rebecca Wingard-Nelson. *Word Problems Using Fractions*. New York, NY: Enslow Publishing, 2017.

WEBSITES

Introduction to Fractions
www.ducksters.com/kidsmath/fractions_introduction.php
Read more about what you need to know about fractions!

What Are Fractions?
www.coolmath4kids.com/math-help/fractions/what-are-fractions
Get some more fraction practice here.

Cyberchase: Fractions
pbskids.org/cyberchase/topics/fractions
Find games, videos, and activities to help you learn fractions.

Publisher's note to educators and parents: Our editors have carefully reviewed these websites to ensure that they are suitable for students. Many websites change frequently, however, and we cannot guarantee that a site's future contents will continue to meet our high standards of quality and educational value. Be advised that students should be closely supervised whenever they access the Internet.

INDEX